Yuri Shunin

Spintronic Nanomemory and Nanosensor Devices

AF135557

Yuri Shunin

Spintronic Nanomemory and Nanosensor Devices

Based on Carbon Nanotube-Fe-Pt Interconnects: Models and Simulations

LAP LAMBERT Academic Publishing

Impressum / Imprint

Bibliografische Information der Deutschen Nationalbibliothek: Die Deutsche Nationalbibliothek verzeichnet diese Publikation in der Deutschen Nationalbibliografie; detaillierte bibliografische Daten sind im Internet über http://dnb.d-nb.de abrufbar.

Alle in diesem Buch genannten Marken und Produktnamen unterliegen warenzeichen-, marken- oder patentrechtlichem Schutz bzw. sind Warenzeichen oder eingetragene Warenzeichen der jeweiligen Inhaber. Die Wiedergabe von Marken, Produktnamen, Gebrauchsnamen, Handelsnamen, Warenbezeichnungen u.s.w. in diesem Werk berechtigt auch ohne besondere Kennzeichnung nicht zu der Annahme, dass solche Namen im Sinne der Warenzeichen- und Markenschutzgesetzgebung als frei zu betrachten wären und daher von jedermann benutzt werden dürften.

Bibliographic information published by the Deutsche Nationalbibliothek: The Deutsche Nationalbibliothek lists this publication in the Deutsche Nationalbibliografie; detailed bibliographic data are available in the Internet at http://dnb.d-nb.de.

Any brand names and product names mentioned in this book are subject to trademark, brand or patent protection and are trademarks or registered trademarks of their respective holders. The use of brand names, product names, common names, trade names, product descriptions etc. even without a particular marking in this work is in no way to be construed to mean that such names may be regarded as unrestricted in respect of trademark and brand protection legislation and could thus be used by anyone.

Coverbild / Cover image: www.ingimage.com

Verlag / Publisher:
LAP LAMBERT Academic Publishing
ist ein Imprint der / is a trademark of
OmniScriptum GmbH & Co. KG
Heinrich-Böcking-Str. 6-8, 66121 Saarbrücken, Deutschland / Germany
Email: info@lap-publishing.com

Herstellung: siehe letzte Seite /
Printed at: see last page
ISBN: 978-3-659-77253-5

Spintronic Nanomemory and Nanosensor Devices Based on Carbon Nanotube-Fe-Pt Interconnects: Models and Simulations

Yuri Shunin[1,3], Stefano Bellucci[2], Yuri Zhukovskii[1], Victor Gopejenko[3], Nataly Burlutskaya[3], Tamara Lobanova-Shunina[4], Federico Micciulla[2] and Aldo Capobianchi[5]

[1] *Institute of Solid State Physics, University of Latvia, Kengaraga Str. 8, LV-1063 Riga, Latvia*
[2] *INFN - Laboratori Nazionali di Frascati, Via Enrico Fermi 40, I-00044, Frascati-Rome, Italy*
[3] *ISMA University, 1 Lomonosov Str. 1, Bld. 6, LV-1019 Riga, Latvia*
[4] *Aeronautics Institute, Riga Technical University, 1 Lomonosov Str. 1, Bld. V, LV-1019 Riga, Latvia*
[5] *Istituto di Struttura della Materia-CNR A.d.R. Roma, Via Salaria Km 29.300 Monterotondo-Rome, Italy*

1

Abstract. The parametrically controlled production of CNTs (carbon nanotubes) with predefined morphologies is a topical technological problem for modern nanoelectronics. The CVD (chemical vapor deposition) technique for SWCNTs (single walled carbon nanotubes) in the presence of various metal nanoparticle catalysts is generally used now. The application of a magnetically stimulated CVD process scheme and catalyst nanoparticles with a strong magnetism promises additional possibilities for the CVD process management and allows expecting a predictable growth of CNTs with set chiralities and diameters. The main attention is focused on the magnetically anisotropic Fe-Pt in $L1_0$ crystallographic phase nanoparticles effect research. The developed theoretical cluster approach based on the multiple scattering and effective medium approximation is used for simulation of fundamental electromagnetic properties in Fe-Pt $L1_0$-CNT interconnects, which are responsible for developing CNTs morphologies. The proposed model of "effective bonds" and the model of magnetic stimulation for growing CNTs morphologies generated on the Fe-Pt nanoparticle surface are applied for the evaluation of the expected CNT chiralities distribution. The model and conditions controlled magnetically, which stimulates CNTs growth in the CVD process, aimed at the predictable SWCNT diameter and chirality and based on Fe-Pt $L1_0$ catalyst are discussed. The possibilities of CNT forest growing on Fe-Pt nanoparticles for magnetic nanomemory are also evaluated. Magnetoresistance phenomenon - giant magnetoresistance and tunneling magnetoresistance - (GMR and TMR) for nanomemory devices based on CNTs of various morphologies (i.e. various chiralities, diameters) which includes metal- and semiconductor-like ones is considered as alternative variants of electromagnetic nanosensing and magnetic nanomemory. Spin transport models are also analyzed.

Keywords: chemical vapor deposition, carbon nanotubes, magnetically controlled growth, Fe-Pt nanodrops-catalysts, CNT growth modelling, magnetoresistance phenomena

Table of Content

1 Introduction

CNTs (Carbon nanotubes) of various chiralities open new wide possibilities for modern nanoelectronics as promising candidates for nanointerconnects in high-speed electronic nanosensoring and nanomemory devices [1-7]. We focus our current study on the implementation of advanced simulation models for a proper description of the fundamental electromagnetic properties (electrical resistance, capacitances and impedances) in contacts between carbon nanotubes of different morphologies and metallic substrates of different nature. We also present the model of magnetically stimulated CNT growth for a special case of Fe-Pt metallic nanoparticles, which have unique magnetic properties. We expect that in the presence of magnetic field, the CNTs growth will be more determined from the point of view of possible CNTs morphologies. Moreover, the creation of a CNT forest based on PtFe nanoparticles provides the possibilities to consider this kind of structure as the basic fragment of nanomemory devices, where information bits are located in nanoparticles and the CNT forest provides the necessary spin transport for reading and recording information.

Thus, in our simulations, we expect to reproduce not only a CNT forest with the predefined morphology but also to develop a prototype of a nanomemory device. The adequate description of CNT chirality [2] is one of the key points for a proper simulation on electric properties of CNT-based nanoelectronic devices. The further development of the cluster approach allowed us to formulate the "effective bonds" model [5] and to carry out a cycle of simulations on electromagnetic properties in various CNTs- and graphene (GNRs - graphene nanoribbons) -based metal interconnects (Me = Fe, Cu, Ag, Pt, Au, Pd, Ni) [3, 8]. We consider that this model serves as a tool for understanding the process of CNT growth adequate to the CVD process.

2 Research Motivation

The main goal of the current research is to understand if there is a relationship between the use of magnetic catalysts and the CVD growth of CNTs, taking into account that most commonly used materials for the growth of CNTs are just Fe, Co and Ni nanoparticles. The nanoparticles of these catalysts are magnetically isotropic. The key question arises: What would happen if one used instead of Fe, Co, and Ni nanoparticles, magnetically anisotropic nanoparticles such as those in the alloy Fe-Pt? The anisotropy of Fe-Pt alloy is due to a spin-orbit coupling of Fe and Pt orbitals. This coupling takes place only if the alloy is formed of alternating planes of Fe and Pt. This particular structure is called L10. In this way, because of the different sizes of the atoms of Fe and Pt, the structure is crushed and atoms of Fe and Pt are close forming a centered tetragonal phase (fct). This approach allows for the coupling and then the magnetic anisotropy with the magnetization axis perpendicular to such planes. It is essential in this case to find conditions to control the growing CNT chirality. We also should take into account a possible substitutional disorder of Fe_xPt_{1-x} alloy, when the stoichiometry number 'x' becomes an additional parameter of CNT growth. No doubt, the diameter of the grown CNT is also an essential parameter, which is evidently pre-defined by the created nanoparticle diameter. But again, we should discuss some limitations in the creation of Fe-Pt nanoparticle sizes in connection with the Fe-Pt melting point. However, Fe-Pt nanoparticles demonstrate an extremely strong coercing field (about, for the phase fct in some special compositions and this promises definite hopes for the controlled CNT growth in the CVD process. We also consider the CVD process of CNT growth as a more predictable one from the point of view of the expected growing CNT parameters (diameter, chirality, morphology of SW or MW CNTs)).

Our experience with the nanotubes synthesis by means of the arc discharge brought about large doubts in the possibilities of CNTs growth control. Two cathodes of pure graphite (99.7%) were taken φ = 6mm (cathode) and φ = 10mm (anode) in diameter. Carbon Black was added to iron – platinum powder, after which the mixture was introduced to complete the filling of the hole. At this point the whole mass was placed in the room. The synthesis was carried out according to the parameters presented here: Current 90 A, Voltage ~ 18 ÷ 20V DC, Vacuum ~ 2×10^{-4} mbar, Gas: He, Work Pressure 600 mbar, arc duration ~ 15 min.

Some filaments also exit from the outer wall of the cathode but in smaller quantities than found on the anode. The residue of the deposit formed on the upper part of the synthesis chamber was collected. It appeared to be quite little compared to other discharges with other catalysts (Figures 1 and 2).

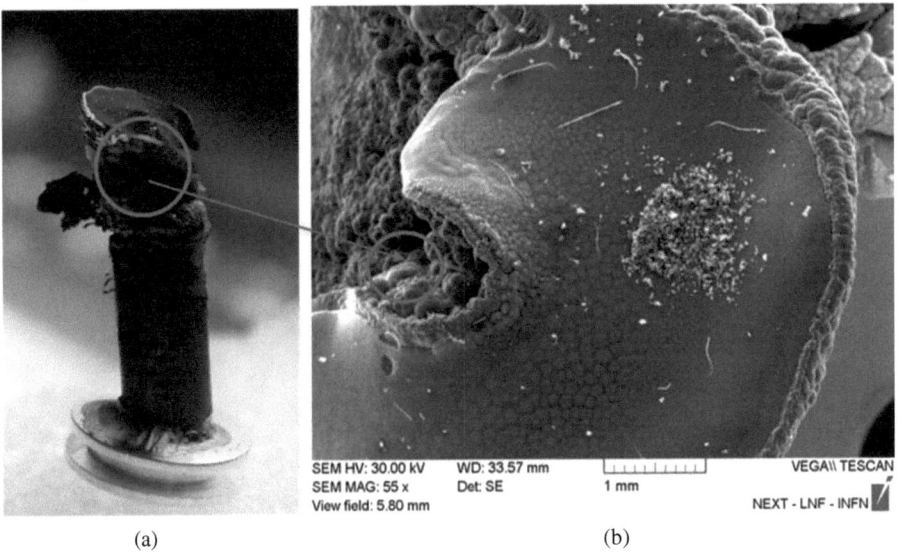

(a)   (b)

Figure 1 (a) Photo of the cathode filaments exiting out after the synthesis, (b) SEM photo of the cathode surface.

The Figure 2 shows that filaments are of different nature - some formats are created from these spheres and others have very fine structures of a few nanometers, probably single-walled nanotubes, covered by these spheres, partly or totally. The question is: What percentage of nanoparticles was originally made or how much iron and platinum and how this underwent a change, during the arc discharge process?

We also take into account the extremely marginal parameters of the Fe-Pt system: Pt has the melting point at T = 2,041.4 K (1,768.2 °C) and the boiling point at T = 4,098 K (3,825 °C).

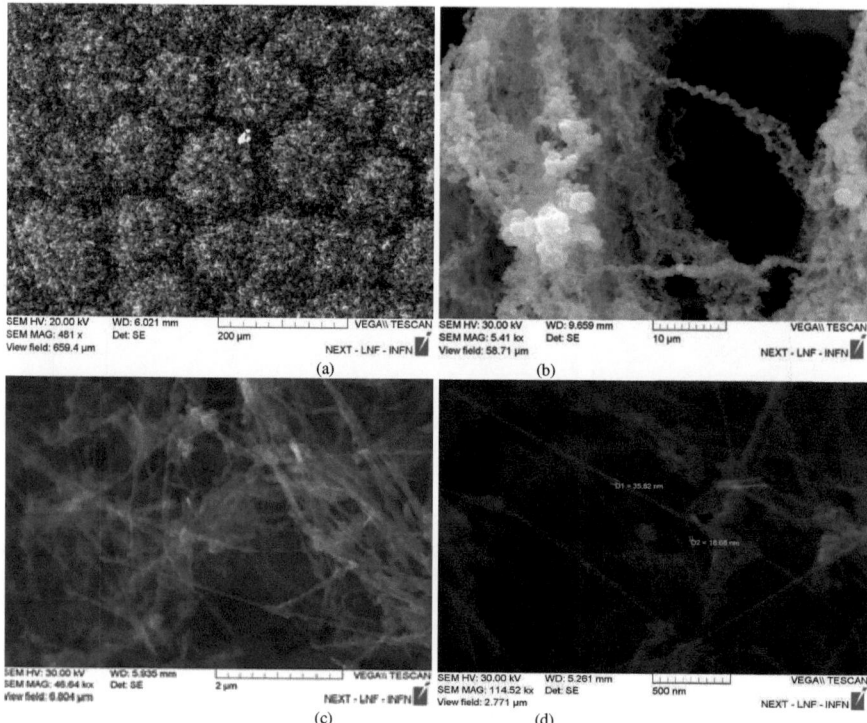

(a)	(b)
(c)	(d)

Figure 2 Evolution view of filaments in the arc discharge process (Bellucci-Capobianchi): (a), (b), (c), (d) [9-12].

There are four major crystal structure modifications for Fe: below 769 °C (Curie temperature) it is α-Fe (α-ferrite) with a body-centered cubic crystal structure and ferromagnetic properties. Ferrite above the critical temperature (769-917 °C) is beta-ferrite (β-Fe) where it is paramagnetic rather than ferromagnetic and it is crystallographically identical to α-Fe. Within the interval of 917-1,394 °C it is γ-Fe (austenite) with acicular cubic crystal structure. At temperatures between 1,394 °C and 1,538 °C, the body-centered cubic crystal structure is a more stable form of delta-ferrite (δ-Fe), the melting point 1,538 °C (1,811 K), and the boiling point at T = 3,273 K (3000 °C). Our earlier research proved that close to the arc the temperature ranges between 4,000 – 6,000 K and there is a subsequent rapid temperature gradient decrease with the distance from it. Filaments are found nearby the arc itself, more precisely, upstream of it. Taking into account the essentially non-steady character of the arc discharge process and very high working temperatures, the CNT growth control is practically impossible and CNT morphologies are non-predictable.

3 CNTs Growth in the Chemical Vapor Deposition Process Based on Metal Nanoparticles

The CVD (chemical vapour deposition) is a highly versatile approach to producing nanotubes and is, perhaps, the most commonly published technique for nanotube growth often used to synthesize CNTs for commercial applications. The process involves decomposition of a carbonaceous precursor at high temperatures under oxygen-free conditions and reduced atmosphere to produce nanotubes.

CVD process begins with previously made supported catalyst material placed in a furnace. A flow of gaseous carbon precursor is introduced for a given time period follows the heating of the furnace till the appropriate reaction temperature. Heating periods can reach several hours. The reaction product is collected from the reactor walls and support surfaces after cooling. There are various experimental conditions for CVD realization in SWCNTs production procedure. Catalyst and carbon precursors, flow rates, reaction temperatures and batch run-times together with support materials are presented when available. There two growth mechanisms in the CVD process: root (base) and tip growth. In root growth, the catalyst particle stays pinned at the support surface and the support – particle interaction must be taken into consideration. In the tip growth the catalyst is lifted off from the support surface during CNT growth due to weak support – catalyst interaction. Evaporation of the catalyst metal may take place during CVD synthesis due to the physical properties of nanosized catalyst particles (nanodrops). The vaporized metal may nucleate and form secondary metal particles away from the surface. The catalytic decomposition reaction may then take place in the gas phase on the surface of the newly formed particles. This product is eventually deposited on the reactor walls.

3.1 CVD PROCESS ANALYSIS

The Chemical Vapour Deposition (CVD) is a highly versatile approach to producing nanotubes and, perhaps, it is the most commonly published technique for nanotube growth often used to synthesize CNTs for commercial applications. The process involves decomposition of a carbonaceous precursor at high temperatures under oxygen-free conditions and reduced atmosphere to produce nanotubes.

The carbonaceous precursor may be a carbon gas, methane or ethylene, for instance, or it can be in the form of a volatile hydrocarbon solvent such as ethanol, which is generally fed into the reactor with some inert carrier gas. In the reaction chamber, the precursor is decomposed in the presence of a catalyst –usually a transition metal such as iron, cobalt or some similar combination of metals. The catalyst can be introduced in a number of different ways.

The substrate needs to have catalyst nanoparticles providing the growth of CNTs where nanoparticles present the place from which nanotubes start growing. Nanoparticles can be composed of lots of different substances (usually, a metal like Fe, Co, Ni Co, Mo, Mn, Pd etc) [13].

The substrate is then heated to the temperature of about 700-800°C under H_2. As the nanoparticles heat up, the hydrogen pulls oxygen out from the nanoparticles, leaving behind metallic nanoparticles.

At the target temperature, a carbon-containing gas such as ethylene (or some others, e.g., alcohol vapor) is introduced. At these temperatures, the gas partly decomposes, giving carbon-containing fragments and other molecules. These fragments and molecules then work their way onto the catalyst nanoparticles, where they stick and then break down into carbon. The carbon dissolves with some probability into or on the nanoparticle. When a critical concentration of carbon is reached in the nanoparticle, the addition of just a little bit more carbon from the vapor causes carbon to precipitate as a crystal.

When carbon crystallizes under the atmospheric pressure (like in CVD), it crystallizes into flat sheets of hexagonally-patterned carbon. But because the nanoparticle is so small and has such a high curvature, when the carbon crystallizes out, it is constrained into a cylindrical shape in the form of a carbon nanotube.

Additional carbon atoms on the nanoparticle provide further nanotube growth, which stops when the carbon-containing gas is turned off if the substrates cool down, or for a number of other reasons that can influence the nanotube or catalyst nanoparticles.

Tuning the appropriated parameters, it is possible to obtain a huge amount of CNTs in a cheap and quick way. Also it is the technique enabling the growth of CNTs with specific characteristics: particular diameter, length and orientation (alignment), and family (Multi or Single Wall). It has been well known for a long time that carbon nanotubes are synthesized by catalytic decomposition of hydrocarbon [14] in the reactor, where metal nanoparticles are presented as a catalyst on a substrate. Catalyst metals mostly used for these purposes are listed in the following Table 1.

Hydrocarbon such as methane adsorbed on the catalytic particle surface, releases carbon during decomposition, which dissolves and diffuses into a metal particle. When a supersaturated state is reached, carbon precipitates in a crystalline tubular form.

It is necessary to emphasize that it will be possible to grow CNTs using a gas, both as a catalyst and as hydrocarbon. If a stream of catalyst particles can be injected into the flowing feedstock, it is possible to produce nanotubes in the gas phase. This approach is amenable for scale-up to large-scale production. Sen et al. [17] first reported such a possibility when they used ferrocene or nickolecene as a source of the transition metal and benzene as a carbon source.

Table 1 Most commonly used catalyst metals for the CVD method [14,19].

Catalyst			Temperature, °C	Carbon source
Metal	*Catalyst type*	*Preparation method*		
Fe	Ultra fine particle	Decomposition of metallocene	1,060	Benzene
	Silica support	Pore impregnation	700	Acetylene
	Zeolite or clay support	Ion exchange	700	Acetylene
	Graphite support	Impregnation	700	Acetylene
	Ultra fine particle	Decomposition of metal carbonyl	800	Acetylene
	Silica support	Sol-gel process	700	Acetylene
Co	Ultra fine particle	Laser etching of Co thin film	1,000	Triazine
	Ultra fine particle	Decomposition of metal carbonyl	800	Acetylene
	Silica support	Pore impregnation	700	Acetylene
	Zeolite or clay support	Ion exchange	700	Acetylene
	Graphite support	Impregnation	700	Acetylene
Ni	Graphite support	Impregnation	700	Acetylene
	Ultra fine particle	Decomposition of Ni(C$_8$H$_{12}$)$_2$	800	Acetylene
Mo	Ultra fine particle	Decomposition of Mo*[1]	800	Acetylene
Mn	Ultra fine particle	Decomposition of metal carbonyl	800	Acetylene
Pd	Ultra fine particle	Decomposition of metal carbonyl	800	Acetylene

[1]Mo*=(NH$_4$)$_{25\pm5}$[Mo$_{154}$(NO)$_{14}$O$_{420}$(OH)$_{28}$(H$_2$O)$_{70}$] 350H$_2$O.

This approach yielded MWNTs, whereas their later work [18] with gas-phase pyrolysis of acetylene using a metallocene yielded SWNTs with diameters around 1 nm [19]. CVD-produced CNTs are curved and have high amount of defects, mainly after the purification in the acid bath; it is needed to remove metal nanoparticles inside of tubes.

The main interests related to iron-containing nanoparticles are focused on their potential applications as high-quality magnetic materials. Thus, a new generation of iron–metal nanoparticles is studied to be used as a catalyst in growing CNTs, with a magnetic nanomaterial inside, making it possible to design a precise carbon nanomagnetic device for drug delivery diagnostics.

For example, iron nanoparticles have been widely used as a catalyst for CVD synthesis of multiwalled carbon nanotubes, [20] while iron-molybdenum can act as a very efficient catalyst for the synthesis of either single-walled or multiwalled carbon nanotubes with the CVD method [21-26].

To prevent the magnetic behaviour of these iron – nano-compounds, it is necessary to grow carbon nanotubes of nanocompounds below the Curie temperature. Typically, the growth temperature is around 600 – 700°C at the atmospheric pressure [27].

The potential applications of carbon nanotubes grown for semiconductor and sensor devices are presented for CMOS industrial applications [28]. The low-temperature growth of vertically aligned carbon nanotubes (CNTs) at high growth rates by a photo-thermal chemical vapour deposition (PTCVD) technique using a Ti/Fe bilayer film as a catalyst is presented in [29]. The bulk growth temperature of the substrate is as low as 370 °C and the growth rate is up to 1.3 μmmin-1, at least eight times faster than the values reported by traditional thermal CVD methods.

It should be recognized that the mechanism of CNTs growth is not obvious enough. Technological growth characteristics may differ in detail, although the concept remains the same. The mechanism of carbon atoms deposition on metal catalyst nanoparticles with the subsequent nucleation of CNTs can be considered one of the most effective and practically important. Crystal (e.g., Si) nanoparticles with a set diameter created on a substrate give a high probability of producing CNTs with the regulated diameter. However, the problem of controlling the chirality of CNTs remains a pressing one. Among more effective catalysts, it is possible to distinguish Pt, Pd, Cu, Ag, Au, Si, SiC, Ge, Al$_2$O$_3$ [30] in addition to the earlier investigated Mg, Ti, Cr, Mn, Fe, Co, Ni, Sn, Pb [31-33].

Table 2 Comparison of catalysts for SWCNTs growth [31-33].

Type	State	Size for SWCNT growth	Catalysis for cracking	Carbon solubility	Growth temperature for ethanol (°C)	Growth rate
Fe, Co, Ni	Liquid/solid	<10 nm	Yes	High	600-950	High
Pt, Pd	Liguid/solid	<5 nm	No?	High	850-950	High
Au, Ag, Cu	Liquid/solid	<5 nm	No	Low	850-950	High
Ge	Liquid/solid	<5 nm	No	Low?	850-950	Low?
Si, SiC	Solid	<5 nm	No	No?	850-950	Low
Al₂O₃	Solid	<5 nm	No	No	850-950	Low

3.2 ADVANTAGES OF CVD

Compared to the discharge and laser-ablation methods, CVD is a simple and economic technique for synthesizing CNTs at low temperatures and ambient pressures. Arc- and laser-grown CNTs are superior to the CVD-grown ones. In yield and purity, CVD has advantages over the arc and laser methods. Considering CNTs in relation to structure or architecture control, CVD is the only answer. CVD is versatile in the sense that it offers harnessing plenty of hydrocarbons in any state (solid, liquid or gas), enables the use of various substrates, and allows CNT growth in a variety of forms, such as powder, thin or thick films, aligned or entangled, straight or coiled nanotubes, or a desired architecture of nanotubes on predefined sites of a patterned substrate. It also offers better control over the growth parameters zone, while carbon crystallization (being an endothermic process) absorbs some heat from the metal precipitation zone. This precise thermal gradient inside the metal particle keeps the process going on. This last advantage is the most essential one [30].Taking only two main CNT parameters, namely, CNT diameter and chirality, the problem cannot be solved well enough. There are two marginal cases of CVD growth.

- Tip-growth model – the catalyst–substrate interaction is weak (metal has an acute contact angle with the substrate); hydrocarbon decomposes on the top surface of the metal; carbon diffuses down through the metal, and CNT precipitates out across the metal bottom, pushing the whole metal particle off the substrate; as long as the top of the metal is open for fresh hydrocarbon decomposition, CNT continues to grow longer and longer; the metal is fully covered with the excess carbon and its catalytic activity ceases and the CNT growth stops.

- Base-growth model - the catalyst–substrate interaction is strong; initial hydrocarbon decomposition and carbon diffusion take place similar to that in the tip-growth case; the CNT precipitation fails to push the metal particle up, so the precipitation is compelled to emerge out of the metal apex; carbon crystallizes out as a hemispherical dome, which then extends up in the form of seamless graphitic cylinder; subsequent hydrocarbon deposition takes place on the lower peripheral surface of the metal, and a dissolved carbon diffuses upward; CNT grows up on the catalyst nanoparticle base.

3.3 CNT PRECURSORS

Most commonly used CNT precursors are methane [34-35], ethylene[18,21] acetylene [36], benzene [21] xylene [37] and carbon monoxide[38].

SWCNTs were first produced from the disproportionation of carbon monoxide at 1200 °C, in the presence of molybdenum nanoparticles [39], and later they were produced from benzene [40] acetylene [18], ethylene [41], methane [42], cyclohexane [43], fullerene [44] by using various catalysts.

It means that the working temperature and pressure of the CVD process can be changed in a wide range of ways and the kinetics of carbon atoms deposition can also be widely varied.

In 2002 the low-temperature synthesis of high-purity SWCNTs from alcohol on Fe–Co-impregnated zeolite support was carried out [45] and since then, ethanol has become the most popular CNT precursor in the CVD method worldwide [46-48].

A special interest in binary catalysts becomes principal now. The unique feature of ethanol is explained by the fact that ethanol-grown CNTs are almost free from amorphous carbon, owing to the etching effect of OH radical [49]. Later, vertically-aligned SWCNTs were also grown on Mo-Co-coated quartz and silicon substrates [50-51]. It has been shown that intermittent supply of acetylene in ethanol CVD significantly assists ethanol in preserving the catalyst activity, thus, enhancing the CNT growth rate [52].

Generally, low-temperature CVD (600-900° C) yields MWCNTs, whereas high-temperature (900–1200 °C) reaction favors SWCNTs growth. It means that SWCNTs have a higher energy of formation (presumably owing to small diameters and the fact that high curvature bears high strain energy). Probably, this is the reason that MWCNTs are easier to grow than SWCNTs.

Recent developments in the nanomaterials synthesis and characterization have enabled many new catalysts for the CNTs growth. Apart from popularly used transition metals (Fe, Co, Ni), a range of other metals (Cu, Pt, Pd, Mn, Mo, Cr, Sn, Au, Mg, Al) have also been successfully used for horizontally-aligned SWCNT growth on quartz substrates [53]. It has been also proposed that the active catalyst is Au–Si alloy with about 80 at% Au [54].

3.4 CNT GROWTH CONTROL

It is a general experience that the catalyst-particle size dictates the tube diameter. The particle size dependence and a model for iron-catalyzed growth of CNTs has been reported in [55] . Metal nanoparticles of the controlled size, pre-synthesized by other reliable techniques, can be used to grow CNTs of the controlled diameter [56].

Influence of a catalyst material and concentration. The additional advantage of using the bimetallic catalyst is that CNTs can be grown at a much lower temperature - 550°C. E.g., the melting point of the mixture of Fe and Co is lower than their individual melting points. Moreover, alloys are known to be better catalysts than pure metals. These trends suggest that tri-metallic catalysts should also give interesting results, though the interpretation of the results would be more complicated.

Influence of temperature. There are investigations of the temperature effect on camphor CVD in a wide range of temperatures 500-1000°C [57]. It was noticed that camphor did not decompose below 500 °C. At 550 °C very short-length tubes emerged from the zeolite pores suggesting that the catalyst activity, and hence the CNT growth rate, was quite low at 550°C. However, the CNT growth abruptly increased at 600 °C, and a profound growth was observed all around the zeolite pores. At 650°C and above, the growth rate was so enormous that hardly a zeolite particle could be located amid nanotubes. The CNT diameter is increased with the growth temperature increase. Very pure CNTs, almost free from metallic impurity, were produced up to 750°C. From 750°C onward, both the diameter and the diameter-distribution range increased drastically. It is supposed that at high temperature, the metal atoms agglomerate into bigger clusters leading to thick CNTs. At 850°C and above, SWCNTs began to take shape alongside with MWCNTs and the volume of SWCNTs increased with the increasing temperature. For instance, at 900 °C, samples of large bundles of SWNTs can be observed. The CVD temperature plays the central role in CNT growth. For a fixed metal concentration, the increasing CVD temperature enlarges the diameter distribution. It should also be noted that MWCNTs and SWCNTs can be selectively grown as a function of CVD temperature if the catalyst concentration is properly optimized [57, 58].

Influence of pressure. For the controlled growth of CNTs by CVD, the vapor pressure of the hydrocarbon in the reaction zone is another very important parameter. For gaseous hydrocarbons, a desired vapor pressure in the CVD reactor can be maintained by a limited gas-flow rate and the controlled suction with a rotary pump [59]. In the case of a liquid hydrocarbon, its vapor pressure is controlled by its heating temperature before it enters the reactor [60]. However, for a solid hydrocarbon such as camphor, it is quite problematic to control its vapor pressure. It becomes a function of three parameters: camphor mass, its vaporization temperature, and the flow rate of argon—the carrier gas. By proper optimization of these three parameters, the influx of camphor vapor to the zeolite bed and its decomposition rate were balanced to a great extent, and a record growth of MWCNTs was achieved at atmospheric pressure by CVD [13]. Pure SWCNTs (free from MWCNTs) have been selectively obtained from camphor CVD at low pressures (10–40 torr) where the camphor vapor pressure is quite in tune with the low metal concentration [61]. It is important to note that the melting point of nanoparticles below 10 nm falls abruptly .E.g., a 5-nm Fe and Co particles can melt at about 850 °C and 640 °C, respectively [17]. It leads to essential limitations in the CVD process of small diameter (< 5nm) CNTs growth.

Schematic representation of the basic steps of SWCNT growth on a metal catalyst is usually simulated in three steps [62]:

1) the diffusion of single C atoms on the surface of a catalyst;

2) the formation of an sp^2 graphene sheet floating on the catalyst surface with edge atoms covalently bonded to the metal;

3) the root incorporation of diffusing single C atoms. It has been shown that carbon atoms diffuse only on the outer surface of the metal cluster. At first, a graphene cap is formed which floats over the metal, while the border atoms of the cap remain anchored to the metal. Subsequently, more C atoms join the border atoms pushing the cap up and, thus, constituting a cylindrical wall.

4 Magnetically Stimulated CNTs CVD Growth on Fe-Pt Catalysts

The key question for a CVD process of CNTs growth is the formation of initial interconnects between substrates (catalyst nanodrops). This step pre-defines the future morphology of the created CNTs. Computer simulation of CNT-metal interconnects is an effective way in formulation of technological recommendations. Our systemic theoretical approach in a wide scale simulations and modelling is based on multiple scattering theory and effective media approximation. This is a reasonable and constructive way to solve a set of marginal condensed media problems in cases of essential structural disordering.

4.1 THEORETICAL APPROACH FOR NANOINTERCONNECT SIMILATIONS: *MULTIPLE SCATTERING THEORY AND EFFECTIVE MEDIUM APPROACH FOR CNT GNR INTERCONNECTS' SIMULATION*

We have developed structural models for CNT-Me and GNR-Me junctions, based on their precise atomistic structures, which take into account the CNT chirality effect and its influence on the interconnect resistance for Me (= Fe, Ni, Cu, Ag, Pd, Pt, Au) and pre-defined CNT (or GNR) geometry. In the simplest cases, the electronic structure for the CNT-Ni interconnects, can be evaluated through the DOS for the C-Me contact considered as a 'disordered alloy'[63]. We have developed more complicated structural models of CNT-metal junctions based on a precise description of their atomistic structures. When estimating the resistance of a junction between a nanotube and a substrate, the main problem has been caused by the influence of the nanotube chirality on the resistance of SW and MW CNT-Me interconnects (Me = Fe, Ni, Cu, Ag, Pd, Pt, Au, Fe), for a pre-defined CNT geometry.

Resistances and impedances as the main electromagnetic parameters are considered as a scattering problem, where the current carriers participate in the transport, according to various mechanisms based on the presence of scattering centers (phonons, charge defects, structural defects, *etc.*), including a pure elastic way, called ballistic. The developed computational procedure [4,64] is based on the construction of cluster potentials and the evaluation of the *S*- and *T*-matrices for scattering and transfer, respectively. It allows us to realize the full-scale electronic structure calculations for condensed matter ('black box'), where **influence** means a set of electronic 'trial' energy-dependent wave functions $\Psi_{in}(\mathbf{r})$ and **response** $\Psi_{out}(\mathbf{r})$ gives sets of scattering amplitudes corresponding to possible scattering channels for any 'trial' energy. This allows us 'to decrypt' the electronic spectra of 'black box' [4].

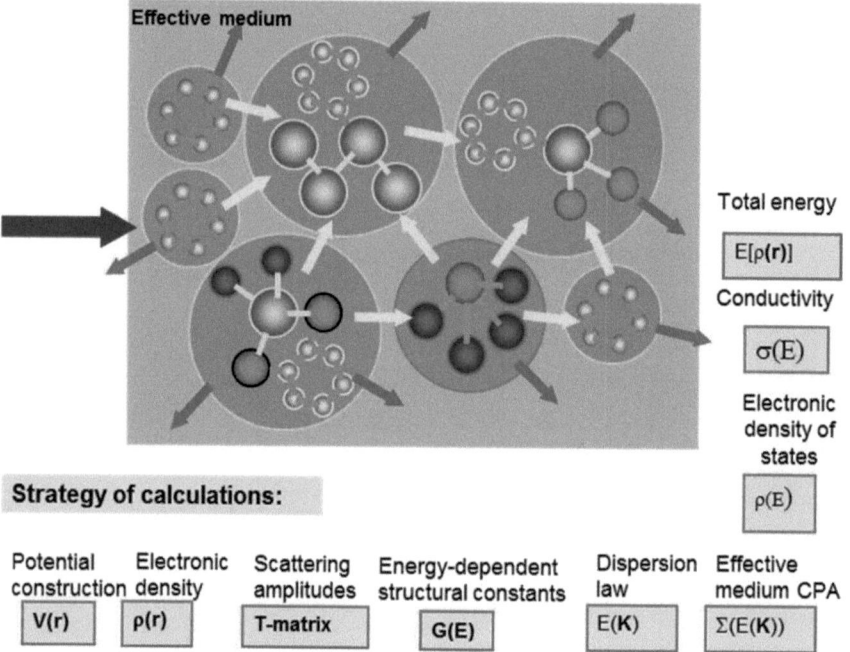

Figure 3 Multiple scattering problem for the system of clusters as multiple scattering model of condensed matter: strategy of calculations on fundamental properties of condensed medium described within the effective media approximation.

We consider a domain where the stationary solutions of the Schrödinger equation are known, and we label them as:

$$\psi_{in}(\mathbf{r}) = \phi_{\mathbf{k}}(\mathbf{r}) = \exp(i\mathbf{k}\mathbf{r}). \tag{1}$$

The scattering of 'trial' waves, in the presence of a potential, yields new stationary solutions labeled as:

$$\psi_{out}(\mathbf{r}) = \psi_{\mathbf{k}}^{(\pm)}(\mathbf{r}) \tag{2}$$

for the modified Schrödinger equation $\hat{H}\psi_{\mathbf{k}}^{(\pm)}(\mathbf{r}) = E\psi_{\mathbf{k}}^{(\pm)}(\mathbf{r})$. An electronic structure calculation is considered here as a scattering problem, where the centers of scattering are identified with the atoms of clusters.

The first step to modeling is the construction of potentials, both atomic and crystalline, which is based on analytical Gaspar's potential of screened atomic nucleus [66] and X_α and $X_{\alpha\beta}$ presentations for the electronic exchange and correlation, using the LDA (Local Density Approximation). Figure 4 shows both atomic and crystalline potentials for carbon as compared to the Hartree-Fock atomic potential. Then, we apply the so-called muffin-tin approximation (*MTA*) for potential models.

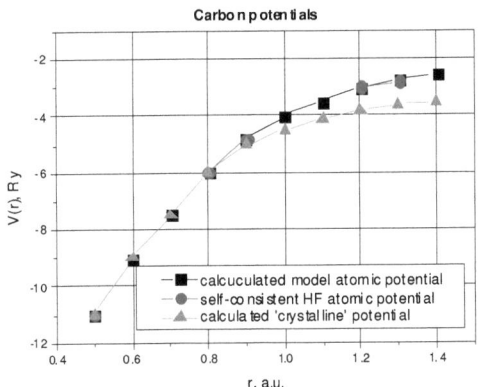

Figure 4 Analytical carbon potentials based on simulation procedure as compared to the results of Hartree-Fock calculations.

To obtain the electronic structure, the calculations on scattering properties are necessary, generally, in the form of *S*- and *T*-matrices (Figure 3). These calculations start with the definition of the initial atomic structure to produce a medium for the solution of the scattering problem for a trial electronic wave [4]. The results of potential modeling and phase shifts in the framework of *MT*-approximation are presented elsewhere [4,64]. The formalism used here for calculations on the electronic structure is based on the CPA [6], the multiple scattering theory and cluster approach [67, 68]. As a *first step*, we postulate the atomic structure at the level of short- and medium-range orders. As a *second step*, we construct a "crystalline" potential and introduce the muffin-tin (*MT*) approach. This is accomplished by using realistic analytical potential functions.

The scattering paradigm for the simplest cases of spherically symmetrical potentials (elastic scattering) looks as follows:

$$\psi(\mathbf{r}) \to e^{ikz} + f(\theta)\frac{e^{ikr}}{r} \quad \text{('liquid metal' model)} \tag{3}$$

and

$$\psi(\mathbf{r}) \to e^{ikz} + f(\theta,\varphi)\frac{e^{ikr}}{r} \quad \text{(spherical cluster model)} \tag{4}$$

Then, the electronic wave scattering problem is solved, and the energy dependence of the scattering properties for isolated *MT* scatterers is obtained in the form of the phase shifts $\delta_{lm}(E)$, and the *T*-matrix of the cluster is found as a whole. The indices l and m arise, as a result of expansions of such functions as Bessel's functions j_l, Hankel's functions h_l and spherical harmonics Y_{lm}.

In general, the modeling of disordered materials represents them as a set of atoms or clusters immersed in an effective medium, with the dispersion $E(\mathbf{K})$ and a complex energy-dependent coherent potential $\Sigma(E)$ found self-consistently in the framework of the CPA. The basic equations of this approach are:

$$\Sigma(E) = V_{eff} + \langle T \rangle (1 + G_{eff} \langle T \rangle)^{-1}, \tag{5}$$

$$G(E) = G_{eff} + G_{eff} \langle T \rangle G_{eff} = \langle G \rangle, \tag{6}$$

$$\langle T(E,\mathbf{K}) \rangle = 0, \tag{7}$$

$$\Sigma(E) = V_{eff}, \tag{8}$$

$$\langle G \rangle = G(E) = G_{eff}, \tag{9}$$

$$N(E) = -(2/\pi)\,ln\{\det\|G(E)\|\}. \tag{10}$$

Here <...> denotes averaging, V_{eff} and G_{eff} are the potential and the Green's function of the effective medium, respectively, $T(E,\mathbf{K})$ the T matrix of the cluster, and $N(E)$ the integral density of the electronic states. Eq. (7) can be rewritten in the form:

$$\langle T(E,\mathbf{K})\rangle = \mathbf{Sp}T(E,\mathbf{K}) = \int_{\Omega_{\mathbf{K}}} \langle \mathbf{K}|T(E,\mathbf{K})|\mathbf{K}\rangle d\Omega_{\mathbf{K}} = 0, \tag{11}$$

where $|\mathbf{K}\rangle = 4\pi\sum_{l,m}(i)^l j_l(kr)Y_{lm}^*(\mathbf{K})Y_{lm}(\mathbf{r})$ is the one-function, $\mathbf{Sp}$ means the calculation of the matrix trace while the integration is performed over all angles of $\mathbf{K}$ inside the volume $\Omega_{\mathbf{K}}$. Eq. (7) enables one to obtain the dispersion relation $E(\mathbf{K})$ of the effective medium. The DOS calculations have been performed using the relation:

$$\rho(E) = \frac{2}{\pi}\int Im\{\mathbf{Sp}G(\mathbf{r},\mathbf{r}',E)\}d\mathbf{r}, \tag{12}$$

where $G(\mathbf{r},\mathbf{r}',E) = \sum_{l,m}Y_{lm}(\mathbf{r})Y_{lm}(\mathbf{r}')G_l(\mathbf{r},\mathbf{r}')$ is the angular expansion of Green function.

The paradigm of the scattering theory and the developed strategy of simulation of CNTs electronic properties use the generalized scattering condition for the low-dimensional atomic structures of the condensed matter:

$$\psi_{\mathbf{k}}^{(\pm)}(\mathbf{r}) \underset{r\to\infty}{\propto} \phi_{\mathbf{k}}(\mathbf{r}) + f_{\mathbf{k}}^{(\pm)}(\Omega)\frac{\exp(\pm ikr)}{r^{\frac{d-1}{2}}}, \tag{13}$$

where Ω describes the integrated space in angular units while superscripts '+' and '-' label the asymptotic behavior in terms of d-dimensional waves:

$$\frac{\partial\sigma_{a\to b}}{\partial\Omega} = \frac{2\pi}{\hbar v}\left|\langle\phi_b|\hat{V}|\psi_a^+\rangle\right|^2 \rho_d(E), \tag{14}$$

where d is the atomic structure dimension.

4.1.1 Calculations of conductivity and resistance

The calculations of conductivity are usually performed using Kubo-Greenwood formula [2]:

$$\sigma_E(\omega) = \frac{\pi \Omega}{4\omega} \int \left[f(E) - f(E + \hbar\omega) \right] \left| D_E \right|^2 \rho(E)\rho(E + \hbar\omega)dE, \tag{15}$$

where ω is a real frequency parameter of Fourier transform for the time-dependent functions, $f(E)$ is Fermi-Dirac distribution function, $D_{E,E'} = \int_\Omega \Psi_{E'}^* \nabla \Psi_E d\mathbf{r}$, where $\Psi_{E(\mathbf{K})} = A exp(i \mathbf{Kr})$ and $\mathbf{K}$ is the complex wave vector of the effective medium. The dispersion function $E(\mathbf{K})$ determines the properties of the wave function $\Psi_{E(\mathbf{K})}$ upon the isoenergy surface in $\mathbf{K}$-space.

For static conductivity ($\omega = 0$ and T = 0 K), Eq. (16) gives the Drude-like formula:

$$\sigma_{E(\mathbf{K})} = \frac{e^2 n^*}{m^*} \tau \quad, \tag{16}$$

where n^* is the effective electron density with a relaxation time $\tau \approx l / v_h$, $l(T)$ is the free path while a heat velocity is $v_h = (3kT/m^*)^{1/2}$. The effective electron mass can be defined using the dispersion law:

$$m^* = (\partial^2 E / \partial K_R^2)^{-1}, \tag{17}$$

where K_R is a modulus of the real part of the vector $\mathbf{K}$.

There are some possibilities to estimate the conductivity in static and frequency regimes taking into account the temperature effects. However, in the case of CNT we must consider not only the diffusive mechanism of conductivity, but also the 'so-called' ballistic one. This is an evident complication in the interpretation of electrical properties of CNTs and the related systems.

4.2 EFFECTIVE BONDS MODEL FOR CNT-Fe-Pt INTERCONNECTS ELECTROMAGNETIC PROPERTIES

These atomistic structures are in compliance with the proposed "effective bonds" model. The "effective bonds" are responsible for mechanical, electronic, magnetic and electrical properties of interconnects. The common consideration of two marginal carbon structures (CNTs and GNRs) is induced by the similar technological problems in respect of these materials for the modern nanoelectronics.

We have developed structural models for CNT-Me and GNR-Me junctions, based on their precise atomistic structures - clusters, which take into account the CNT chirality effect and its influence on the interconnect resistance for Me (= Fe, Ni, Cu, Ag, Pd, Pt, Au) as well as the pre-defined CNT (or GNR) geometry. These atomistic structures are in compliance with the proposed 'effective bonds' model. The 'effective bonds' are responsible for mechanical, electronic, magnetic and electrical properties of interconnects. The common consideration of two marginal carbon structures (CNTs and GNRs) is induced by the similar technological problems in respect of these materials for the modern nanoelectronics.

The chirality (m,n) is simulated by the corresponding orientation of carbon rings within the scattering medium (Figure 5).

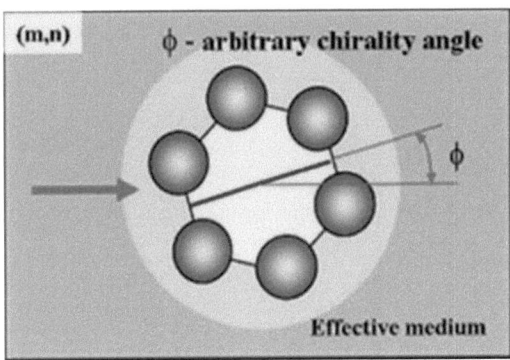

Figure 5 Modeling of chirality: carbon ring rotation within CNT and GNR.

The most problematic areas for the proper simulation are CNT-Me and GNR-Me *end to* junctions (Figure 6), where the atomic structural disorder is observed and the conductivity mechanism is changed.

The influence of chirality on resistance in the vicinity of interconnect depends on the number of statistically realized bonds between the CNT (GNR) and the metal contact (*e.g.*, Fe, Ni, Cu, Au, Ag, Pd, Pt).

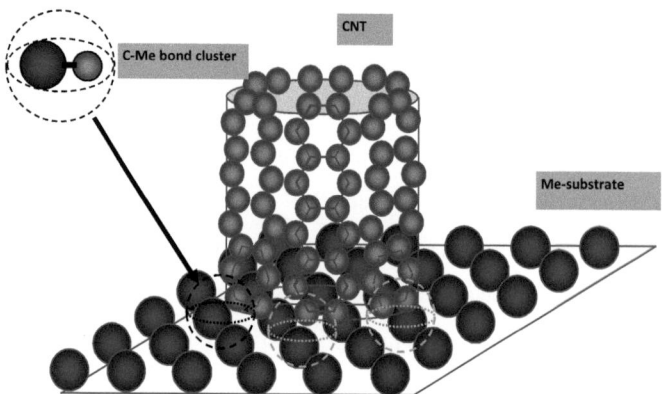

Figure 6 CNT-Me interconnect formation model.

Our main interest in this particular case is focused entirely on CNT-Me interconnects, when the modelling of CNTs growth on metal surfaces is combined with the controlled electromagnetic properties in the interconnect area. To ensure a predictable interconnect morphology, the permanent magnetic field is used that accompanies the growth, and magnetic drops are used as growth catalysts that provide for the special composition of Fe-Pt with unique magnetic properties. The results of our simulations show that interconnects resistance and the number of effective bonds can be considered as indicators of chirality. Figures 7-11 demonstrate the numbers of effective bonds via CNT diameters, chirality angles, CNT-Me interconnects resistances and impedances. It means that resistances and the number of effective bonds in the interconnect space are indicators of CNTs morphology.

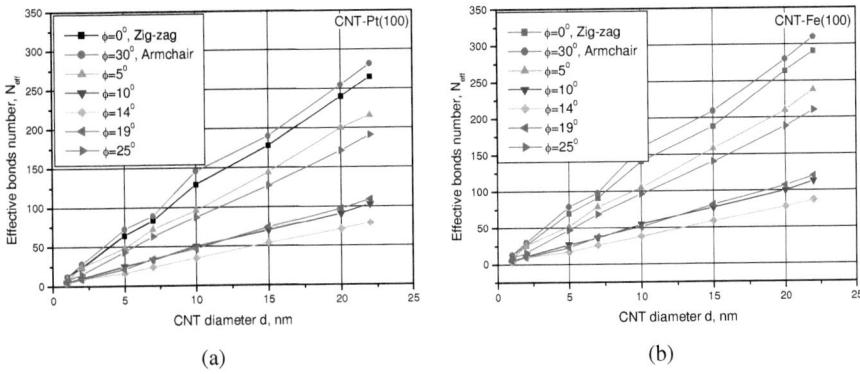

Figure 7 Effective bonds number via CNT diameter: (a) CNT-Pt and (b) CNT-Fe.

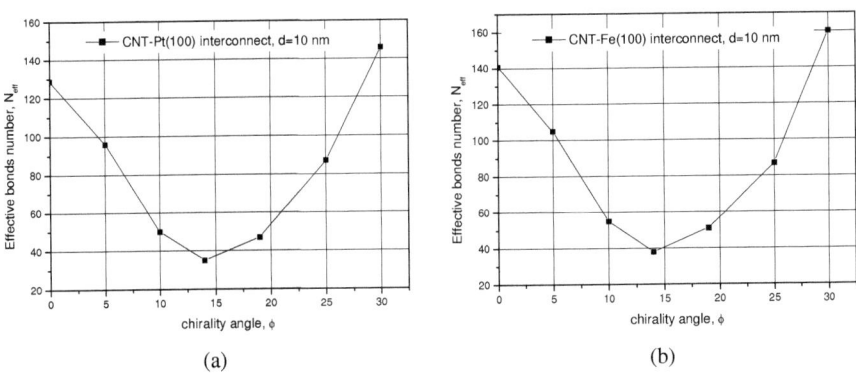

Figure 8 Effective bonds number via CNT chirality for the diameter d = 10 nm: (a) CNT-Pt and (b) CNT-Fe.

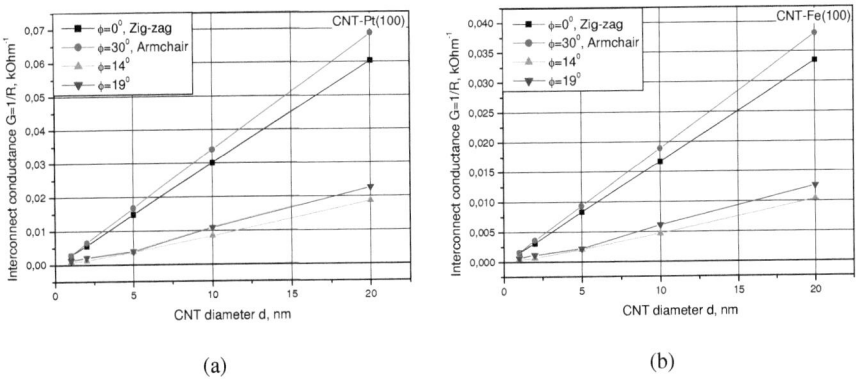

Figure 9 Interconnect conductance via CNT diameter: (a) CNT-Pt and (b) CNT-Fe.

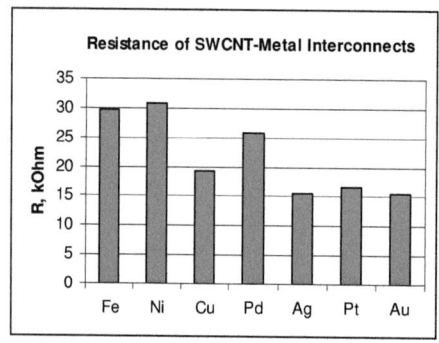

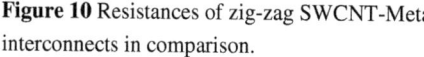

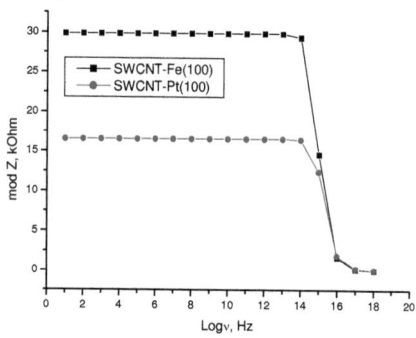

Figure 10 Resistances of zig-zag SWCNT-Metal interconnects in comparison.

Figure 11 Impedances of SWCNT-Fe and -Pt interconnects in comparison.

4.3 CNT-Fe$_x$Pt$_{1-x}$ INTERCONNECT FORMATION

There is a relation between the use of magnetic catalysts and the CVD growth of CNTs determining the most commonly used materials for the CNT growth - Fe, Co and Ni. The nanoparticles of the latter catalysts are magnetically isotropic. Magnetism of the particles-catalysts brings order into the process of CVD growth.

Actual class of Fe-Pt nanodrops as possible catalysts in a CVD process has the diameter ranging from 5 to 30 nm. The stable and predictable creation of definite sizes of nanodrops is a basis for the output growth of CNTs with the expected morphology. Their magnetic properties are essentially changed in dependence on their size. Magnetization orientation of nanodrops in a Fe-Pt system is regulated in competition of the heat factor and collective magnetization of the nanodrop ferromagnetic state. Magnetization of nanodrops becomes non-stable for small sizes. This also means additional technological complications in magnetically controlled CMTs CVD growth. Fe-Pt nanoparticles containing a near-equal atomic percentage of Fe and Pt are an important class of magnetic nanomaterials. They are known to have a chemically disordered face-centered cubic (fcc) structure or a chemically ordered face-centered tetragonal (*fct*) structure.

Additional possibilities in controlling the process of CVD growth open up the ways to use magnetically anisotropic nanoparticles such as those in the alloys with a different substitutional disorder (*e.g.*, Fe_xPt_{1-x}), to manage the CVD process with the formation of the predefined CNT chiralities. Besides, the stimulation of the process by means of the external magnetic field activates magnetic moments of the catalyst and the deposited carbon atoms. This also means that we can control the number of effective bonds inside interconnects. We pay a special attention to Pt and Fe substrates as possible elements for nanoelectronic and nanomagnetic devices considering interconnects fundamental properties and magnetically stimulated nanoprocesses on Fe-Pt $L1_0$ nanoparticles [10-12, 69].

4.3.1 Magnetic properties of Fe-Pt alloys

The unique magnetic properties of Fe-Pt alloys - is an open field for research to correlate different CNT techniques in terms of the catalyst role in entirely different range of temperatures and pressures. Carbon nanotubes grow using bimetallic nanoparticles Fe-Pt as a catalyst [70-73]. Since the mid-1930s Fe-Pt alloys have been known to exhibit high coercivities due to high magnetocrystalline anisotropy of the $L1_0$ FePt phase but their high cost prevented these alloys from widespread applications in the past. In Fe-Pt alloys, both Fe and Pt atoms carry a magnetic moment: the induced magnetic moment on the Pt sites and the enhanced magnetic moment on the Fe sites. A wide variety of the magnetic structure types in the Fe-Pt alloys is evidently the consequence of various atomic configurations around Fe atoms, which, in turn, has a considerable effect on the electronic structure of these alloys (see, Figure 12, Table 3, and Eutetic Phase Diagram of Fe-Pt alloy (http://www.himikatus.ru/art/phase-diagr1/Fe-Pt.php)).

Table 3 Specific magnetic structures of Fe-Pt systems.

Magnetic structure	Behaviour	Curie temperature
L1$_2$ Fe$_3$Pt	Ferromagnetic	410 K
L1$_0$ FePt	Ferromagnetic	750 K
L1$_2$ FePt$_3$	Paramagnetic	273 K

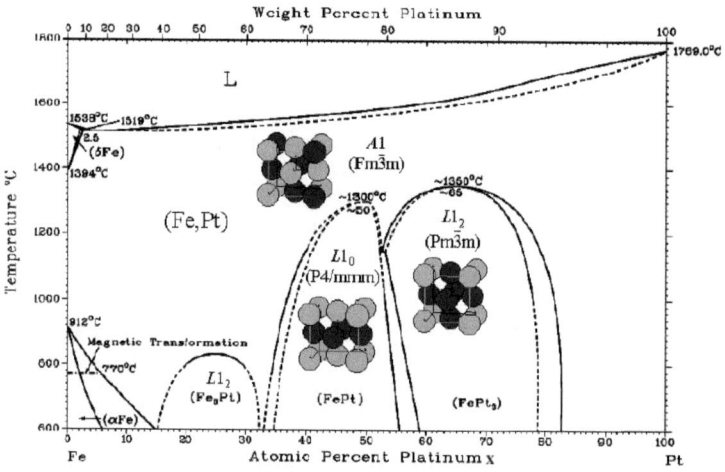

Figure 12 Eutetic Phase Diagram of Fe-Pt alloy:
(http://www.himikatus.ru/art/phase-diagr1/Fe-Pt.php).

To obtain the Fe-Pt alloy in L1$_0$ phase we use a new chemical synthesis strategy reported in details in our previous works [10-12, 69]. By planetary ball milling of the nanocrystals of a particular precursor [Fe(H$_2$O)$_6$]PtCl$_6$ and NaCl were grounded and then annealing at 400 °C (300 °C less than typical used temperature) in reductive atmosphere, After washing we obtain FePt L1$_0$ NPs with selected size as function of fraction of [Fe(H$_2$O)$_6$]PtCl$_6$ and NaCl used. By varying the precursor ratio we are able to avoid the coalescence phenomena and obtain single crystal NPs with the size around 6 nm, not agglomerated.

4.3.2 Magnetically stimulated CNTs growth

The formation of the initial optimal perimeter for C-Metal (Fe-Pt) bonds is a synergetic process with a minimal free energy (see Figure 13).

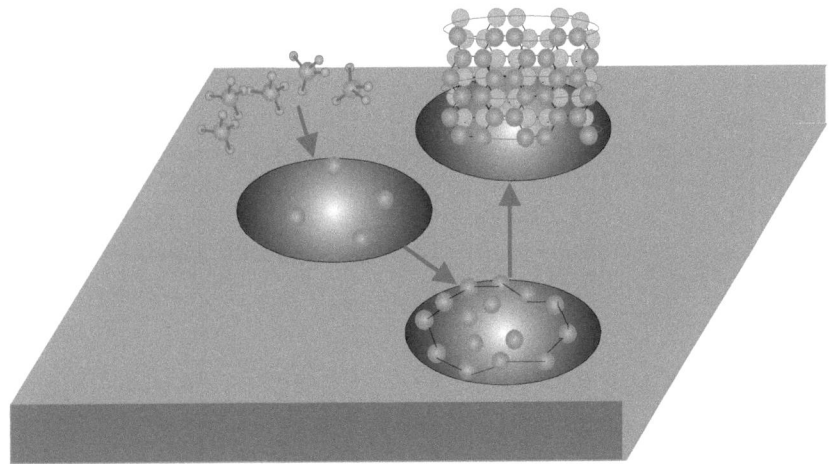

Figure 13 A fragment of the CNT CVD process growth on nanoparticle surfaces.

The nanoparticle diameter determines with a certain error the diameter of a CNT. The number of effective bonds defines the morphology of the future CNT (See Figure 14, arm-chair, chiral, zig-zag CNTs,) in terms of chirality. Obviously, there is a considerable uncertainty in the morphology of the future CNTs, owing to sporadic thermal dynamics of the deposited carbon atoms.

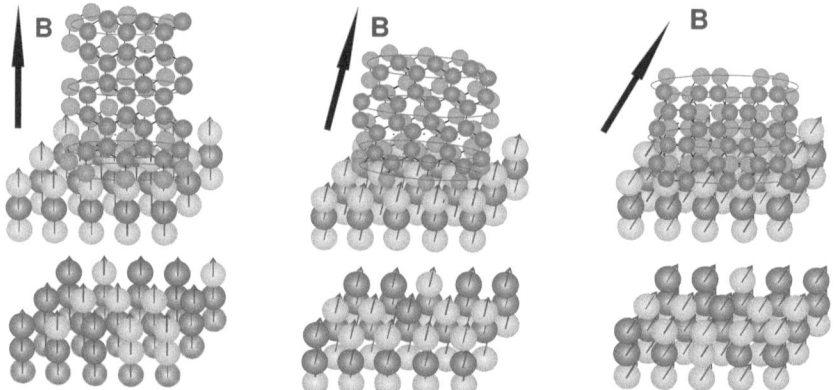

Figure 14 Magnetically stimulated orientation of magnetic moments of atoms in model interconnects during the CNTs controlled growth with the expected chiralities under the directed magnetic field B with the orientation angle $\phi = \Theta_B$: (a) arm-chair CNT, magnetic field **B** with the orientation angle Θ_B, where **B** is the magnetic induction and the chirality angle **is** $\phi = 0°$; (b) chiral CNT, $0 < \phi < 30°$ and (c) zig-zag CNT, $\phi = 30°$.

According to our model, the process can be considerably streamlined using even minor diamagnetic properties of carbon atoms at the expense of magnetic field and strong induced ferromagnetism of the nanoparticle. It becomes evident that the nanoparticle composition Fe-Pt and its atomic structure (including the short-range order) play a primary role in the process of CNT growth stimulated by the magnetic field of CVD.

Nucleation in the process of CNTs growth in cases of the ordered Fe-Pt nanoparticles is more stable and has principal advantages in relation to the controlled CNTs growth compared with the cases of any kind of anisotropic Fe-Pt nanocatalysts [74].

The availability of substitutional disorder and anisotropy in Fe-Pt nanodrops used for CNT CVD controlled growth is a negative factor for the production of CNTs with predictable morphologies. There are some essential negative features of disorder in Fe-Pt nanodrops for the CVD process:

1) C atoms chemical binding preferences (C-Fe bonds are stronger than C-Pt ones);

2) A higher structural anisotropy leads to higher relaxation times of a nanodrop structure [75];

3) Disordering effects in Fe-Pt lead to sporadic magnetization phenomena.

Thus, it is important to control the process of formation of nanodrop structures corresponding to $L1_0$ (Fe-Pt) and $L1_2$ (Fe$_3$Pt) systems with reliable magnetism and adequate cooling rates, when relaxation processes in nanodrops are accomplished.

As it has already been mentioned, Fe-Pt nanoparticles containing a near-equal atomic percentage of Fe and Pt are an important class of magnetic nanomaterials. They are known to have a chemically disordered face-centered cubic (*fcc*) structure or a chemically ordered face-centered tetragonal (*fct*) structure.

4.3.3 Model of CVD CNT growth with the probabilistically pre-defined morphology

Magnetically stimulated CNT CVD model formulation looks as follows (see Figures 13 and 14):

- Effective bonds are responsible for the future chirality of CNT in the CVD processes.

- The permanent magnetic field provides the alignment of CVD growth.

- The binding with the Fe-Pt nanoparticle has a probabilistic character; the probability of binding of C atoms with Fe atoms is preferable as compared to Pt atoms.

- The composition of Fe-Pt atoms in a nanoparticle and the atomic structure ordering are essential for chirality of the obtained CNTs.

- The diameter of the future CNT is correlated with the nanoparticle size.

- The direction of the magnetic field stabilizes the direction of CNT forest growth and is a tool of the growth control.

- The ordering of a nanoparticle atomic structure, as well as a substitutional disorder or arbitrary disorder of a nanoparticle, is responsible for the magnetic state of the nanoparticle, which is essential for the possible future nanomemory devices.

Even a very strong magnetic field is not able to suppress thermal fluctuations of magnetic moments in carbon atoms. It is a different thing, when carbon diamagnetic atoms appear on the surface of a catalyst in the ferromagnetic condition where magnetism induced by the external field can be very strong and is able to correct the behavior of carbon atoms by suppressing thermal fluctuations.

At the same time, the growth control over chiral and non-chiral nanotubes essentially depends on stoichiometric composition of Fe-Pt nanoparticles. The beginning of the nucleation process with the growth of nanotubes might be connected with stochastic fluctuations of the magnetic moment in a carbon atom relative to the direction of the local magnetic field in a nanoparticle. Distribution of the fluctuation angle obeys the Gaussian law:

$$f(\theta) = \frac{1}{\sigma\sqrt{2\pi}} \cdot \exp\left(-\frac{(\theta - \theta_B)^2}{2\sigma^2}\right), \tag{18}$$

where σ^2 is the angular dispersion of thermal fluctuations of the magnetic moment angle of a carbon atom. To evaluate this dispersion, the potential energy change of the magnetic moment under the influence of the thermal energy should be evaluated: $\mu_C B(1 - \cos\theta_T) \approx k_B T_{CVD}$, where μ_C is the induced magnetic moment of a carbon atom $\mu_C = 1.25\mu_B$ (see evaluations in [73], $\mu_B = 5.788 \cdot 10^{-5} \ eV/T$, $\mathbf{B}$ is the magnetic induction of the catalyst surface, $\theta_T = \theta - \theta_B$ T_{CVD} - is the operating temperature of the CVD process, $k_B = 8{,}617 \ 3324(78) \cdot 10^{-5} \ eV/K$- is the Boltzmann constant. Hence $2\sin^2\frac{\theta_T}{2} \approx \frac{k_B T_{CVD}}{\mu_C B}$. Taking into consideration one of the main problems of the nanotubes growth control – the chirality control – it is necessary to seek for the small

fluctuation angle θ. Then, $\sigma^2 = \theta_T^2 = \dfrac{2k_B T_{CVD}}{\mu_c B}$.

The condition of the small fluctuation angle (e.g., $<10°$) at a certain temperature of the CVD process imposes limitations on the values of the demanded magnetic induction **B**.

Taking into account the ratio between the chirality angle and the direction of the magnetic field $\phi = \theta_B$, Figure 15 displays the predictable scattering of chiralities for nanotubes of approximately the same diameter.

We are also able to evaluate the necessary value of the magnetic field **B** providing the expected chirality angles scattering, e.g. $\sigma = 0.2$ (approximately $12°$) leads to the

B evaluation for the CVD process temperature $T_{CVD} = 700°C$ as

$B = \dfrac{2k_B T_{CVD}}{\sigma^2 \mu_c} \approx \dfrac{16755}{\sigma^2 \mu_c} \approx 57895 \approx 6 \cdot 10^4 \; Tesla.$ For small angle dispersions

$\sigma^2 = \theta_T^2 = \dfrac{2k_B T_{CVD}}{\mu_c B}$ the high local magnetic field on the nanoparticle surface

is necessary. The result also strongly depends on the carbon atom magnetic moment μ_c .

Speaking about the possible errors in diameters of growing nanotubes, their evaluation from beneath is defined by the minimal variations in parameters of the chirality vector $\vec{c} = (n, m)$ Δn and Δm, which are equal 1.

Taking into consideration the formula for calculating the diameter of CNT:

$d = \dfrac{\sqrt{3}a}{\pi} \sqrt{m^2 + n^2 + mn}$, where $\alpha = 0,142$ nm is the distance between the neighboring carbon atoms in the graphite plane.

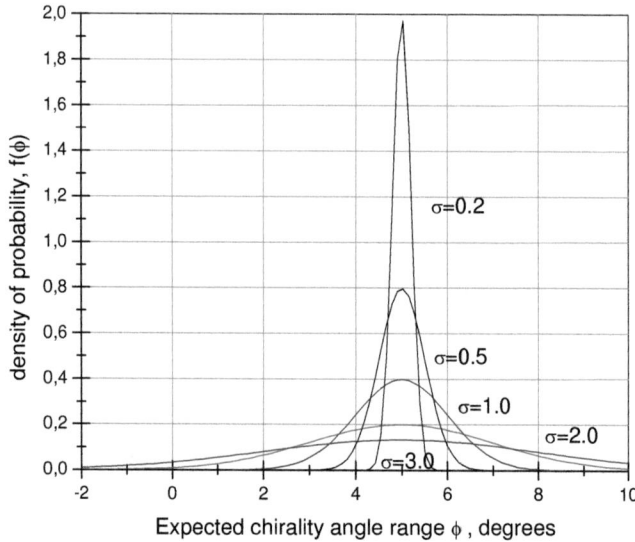

Figure 15 The predictable scattering of chiralities for nanotubes of approximately similar diameters.

The relationship between chirality indices (n and m) and the angle ϕ is given by the ratio:

$$\sin \phi = \frac{m\sqrt{3}}{2\sqrt{m^2 + n^2 + mn}}. \qquad (19)$$

Then the minimum relative error in the diameter of CNT is possible to define as:

$$\varepsilon_d = \frac{\Delta d}{d} \approx \frac{\sqrt{5m^2 + 5n^2 + 8mn}}{2(n^2 + m^2 + mn)}. \qquad (20)$$

Particularly, in the case of arm-chair CNT ($m=0$) $\varepsilon_d \approx \dfrac{\sqrt{5}}{2n}$, in the case of zig-zag

CNT ($m=n$): $\varepsilon_d \approx \dfrac{1}{2\sqrt{2n}}$.

Thus, the errors in the diameter of growing CNTs are incorporated in discrete morphological properties. But these minimum estimates are only reinforced, given the obvious errors in the size of catalyst nanoparticles are taken into consideration. The perfect picture of the magnetically stimulated CVD process for growing CNTs can be presented as a CNT forest (see Figure 16). Such a system of nanotubes can also be considered as a prototype of the magnetic memory, where ferromagnetic nanoparticles serve as cells of the magnetic memory – that is, ferromagnetic contacts are controlled by spin pulses, the transport of which is provided by nanotubes.

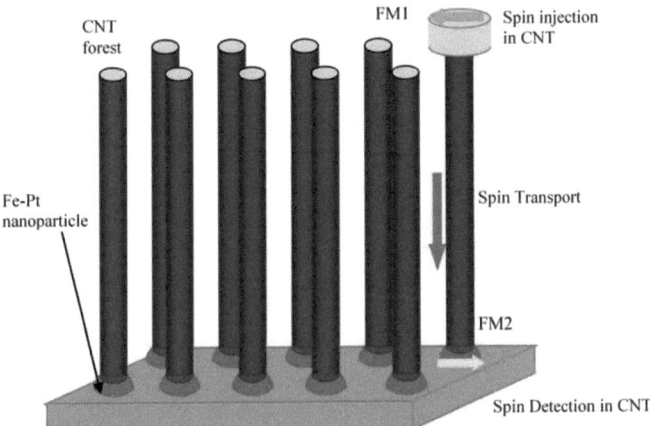

Figure 16 CNTs forest is grown on the Fe-Pt nanoparticles of the predefined radius as a fragment of magnetic nanomemory device, which can be realized if spin injection, spin transport and spin detection (spin recording) are provided.

5 Spintronics and Magnetoresistance Phenomenon for Nanomemory Devices

5.1 SPIN VALVES CONCEPTIONS

Nanocarbon-based (e.g. CNT or graphene) spintronics is well-predicted theoretically. The main point of technological problem is the spin current injection efficiency into carbon nanotubes by using ferromagnetic electrodes. The next point of attention is spin transport experiments to obtain evaluations of digital information processing rates. A typical spintronic device is the so-called spin valve. The typical spin valve device contains two ferromagnetic electrodes (one ferromagnetic electrode for spin current injection, and another for spin signal detection) and non-magnetic material responsible for spin transport. For practical implementation, the spin signal should be large enough. A large spin signal can be obtained in a weak spin relaxation system, where spin can travel a long distance without being scattered in ballistic regimes. Nanocarbon-based materials (e.g., CMTs) due to long spin relaxation distances look as the best candidates in comparison with the nonmagnetic metals and semiconductors. The spin current is propagated for long macroscopic distances due to some spin correlation. These nanocarbon-based materials are expected to be beneficial for implementation of large scale spintronic logic circuits, quantum computing, fast data processing, magnetic hard disk drivers and dynamical nanomemory devices.

Electron transfer in spintronic devices depends on the spin polarization of the current. Such devices include a method of generating a spin-polarized current based on interconnects of nonmagnetic materials with ferromagnetic ones. In our case, these are Fe-Pt compositions.

EDOS spin splitting on the Fermi level defines the efficiency of spintronic devices. In accordance with Stoner model the band structure of transition metals depends on spin orientation (up and down) [85]. The theoretical basis for considering spintronic devices is a magnetoresistance phenomenon. Namely, there are two variants of magnetoresistance - giant magnetoresistance (GMR) and tunneling magnetoresistance (TMR).

Electronic transport through CNTs depends on the contacts with electrodes. For short CNTs at low temperatures we should also pay attention to Kondo effect - the formation of a many-body dynamical singlet between a localized spin and delocalized conduction electrons of electrodes [87]. It should be borne in mind that carbon nanotubes possess in addition to spin also orbital degeneracy. The increase of degeneracy corresponds to the enhancement of Kondo temperature, which is important for potential applications.

Giant magnetoresistance is a quantum mechanical effect observed in thin metal films consisting of alternating ferromagnetic and non-magnetic conductive layers. The effect is a substantial change in the electrical resistance of such a structure when changing the relative direction of magnetization of adjacent magnetic layers. The direction of magnetization can be controlled, for example, by applying an external magnetic field. The effect is based on the scattering of electrons depending on the spin direction.

The GMR was discovered in 1988 (2007 Nobel Prize, Albert Fert and Peter Grünberg) as a large change in resistance of magnetic Fe/Cr multilayer in the presence of an applied magnetic field [76]. Soon after GMR was discovered in Fe/Cr/Fe trilayers [77]. As it was shown later, the effect can be obtained in trilayers having other magnetic materials such as Co [78].

Such trilayer structures, i.e. sandwiches of two ferromagnetic metals separated by a thin spacer layer of normal metal (Figure 17) are of great industrial importance. They are called spin-valves and are used as magnetic field sensors. The resistance of the device is dependent on the relative magnetization orientation of the ferromagnets. It is R_P when the magnetizations are parallel and RA when they are antiparallel. The GMR ratio is defined as $GNR = \dfrac{R_A - R_P}{R_P}$.

The TMR is a quantum mechanical effect manifests itself when a current flows between the two ferromagnetic layers separated by a thin (about 1 nm) dielectric layer.

The total resistance of the device, the current which flows due to the tunneling effect depends on the relative orientation of the two fields, the magnetization of the magnetic layers. Resistance above at the perpendicular magnetization layer. The effect of the tunnel magnetic resistance similar to the effect of a giant magnetic resistance, but in it, instead of a non-magnetic metal layer is used insulating layer of the tunnel barrier.

The TMR was discovered in 1975 by M. Jullière [79] in a device that consisted of two Fe films separated by Ge. It was first in 1995 when room temperature TMR was discovered by J. S. Moodera [80] and caused a great interest in spintronics after GMR was discovered in 1988 [76]. Our idea is that the same effect can be reached by introducing into the N space semiconductor CNTs.

The TMR signal operates in the same way as the GMR: $TNR = \dfrac{R_A - R_P}{R_P}$, where R_P (= $1/G_P$) and R_A (= $1/G_A$) are the resistances (conductances) of the device for parallel and antiparallel orientations respectively of the ferromagnets magnetization. However, the nature of TMR is different than that of the GMR.

The resistance for the antiparallel magnetization is normally higher than for the parallel one. At the applied zero magnetic field, the relative orientation of the magnetization is governed by the exchange coupling between the ferromagnetic layers. The sign and size of the interlayer exchange coupling is dependent on the thickness of the nonmagnetic spacer. The coupling can thus be ferromagnetic and antiferromagnetic dependent on the spacer thickness [81, 82].

5.2 SPIN TRANSPORT IN FM-CMT-FM SYSTEMS

It has been found that carbon nanotubes can acquire magnetic properties when brought into contact with ferromagnetic materials. This property of carbon nanotubes has previously been predicted theoretically by M Ferreira and S Santivo. The effect is based on the exchange of polarized electrons between the ferromagnetic material and nanotubes. The main difficulty of the experiment was in detecting the weak magnetic moment of the nanotubes on the background of a strong magnetic moment of the ferromagnetic sample. The sample used in the experiment was a thin film of cobalt or iron oxide and was uniformly magnetized strictly in one direction.

With the help of a magnetic force microscope, weak perturbing magnetic fields were revealed produced by nanotubes on the surface of the sample. According to the measurements, the magnetization of nanotubes is 0.1 Bohr magneton per one carbon atom, which is in contact with the film. The control experiments demonstrated that carbon nanotubes in contact with nonmagnetic materials such as Si or Au do not possess magnetism [88].

The general approach is to examine the chain of weakly interacting CNT spins in contact with the two ferromagnetic metals (FM). Let us consider Hamiltonian **H** of the system of localized spins whose interaction is described by the model of Ising or Heisenberg:

$$
\begin{aligned}
\mathrm{H} &= -\frac{1}{2}\sum_{l,l'} J_{l,l'}\left[\alpha(S_l^x S_{l'}^x + S_l^y S_{l'}^y) + S_l^z S_{l'}^z\right] - \mu\sum_l S_l \boldsymbol{H} = \\
&= -\frac{1}{2}\sum_{l,l'} J_{l,l'}\left[\alpha(S_l^x S_{l'}^x + S_l^y S_{l'}^y) + S_l^z S_{l'}^z\right] - \mu H \sum_l S_l^z,
\end{aligned}
\tag{21}
$$

where s_l - the spin localized in the lattice site **l**, **H** - external magnetic field, μ - effective magneton, $J_{l,l'} = J(\boldsymbol{R}_{l,l'})$ - the energy of the exchange interaction (the exchange integral) of the nearest neighbors in the lattice. If $\alpha = 1$, Hamiltonian corresponds to the Heisenberg model, with $\alpha = 0$ – to the Ising model. The magnetic disordering in the system of CNT- type is due to the weak magnetism of diamagnetic carbon atoms, where the only essential thing is the level of magnetic interaction of the short-range order. In this case, a regular atomic structure of CNT is accompanied by a magnetic disorder of diamagnetic-liquids type. The correlation of the magnetic moments of CNTs can be ensured by a spin injection at the expense of the controlled magnetic field in the area of ferromagnetic contacts. At the same time, the initial values $J_{l,l'}$ of the energy of the magnetic interaction till the controlled magnetic pulse have random values. The weak correlation between adjacent carbon atoms in the atomic structure of CNT is also supported by the excitation of FM-CMT-FM. Accordingly, in (21) summation is assumed only considering the efficiency degree of unbroken bonds.

In the Heisenberg model, the node approximation with the *l*-th node of the ideal crystal is associated with a localized magnetic moment proportional to the localized spin variable S_l.

In the case of paramagnetic materials, in the absence of an external magnetic field, S_l are oriented randomly from node to node and form a system with magnetic disorder. Diamagnetic materials exhibit their magnetic properties due to the induced magnetic moments in the presence of a magnetic field. However, the magnetic moments are very small, and their correlated behavior is characterized only by a short-range order. Therefore, in terms of spin transfer, paramagnetic and diamagnetic materials are quite similar.

In the case of cooperative spin interactions at low temperatures, magnetic order appears. It is characteristic for ferromagnets, antiferromagnets and ferrites. However, the order disappears at high temperatures and magnetic system itself is transformed into a paramagnetic with a certain level of magnetic disorder. Considering that S_l – is a vector quantity, the types of disorder may be different. Of course, the thermal fluctuations significantly stimulate spin disordering.

Using the Ising model in its pure form (when $S_z = \pm 1$) for the systems such as FM-CMT-FM is not correct, because each node has three spin variables - S_x S_y and S_z, which can take random values, and are connected by the relation: $S^2 = S_x^2 + S_y^2 + S_z^2$ and quantized by being inherently quasicontinuous. This is due to the boundary conditions on the FM-CNT interconnects. Really used constructive models of spin coupling (Ising model, classical and quantum Heisenberg model) for the system under investigation require further modifications. For example, in the near-field approximation, it is possible to use one parameter of the exchange interaction J in the model Hamiltonian:

$$H = -\frac{1}{2}\sum_{l,l'} J S_l S_{l'} - \mu H \sum_l S_l. \tag{22}$$

Further steps may be associated with a version of Heisenberg magnetic system and the corresponding Hamiltonian:

$$H = -\frac{1}{2}\sum_{l,l'} \xi_l \xi_{l'} J(R_{ll'}) S_l S_{l'} - \mu \sum_l \xi_l S_l H, \tag{23}$$

where ξ_l - a random value from 0 to 1, which can be adjusted by the temperature factor, that is $\xi \square \exp(-W/kT)$, where W - is m the magnetic interaction energy of the induced magnetic dipole with an external magnetic field.

Obviously, this type of complex Hamiltonian needs to be simplified. For example, we can neglect the spin deviations along the z axis, and consider only the transverse deviations introducing spin variables such as: $S^{\pm} = S_l^x \pm iS_l^y$.

The linearized equations of motion look as follows:

$$i\hbar \frac{dS_k^-}{dt} = 2SJ \sum_{k'} \left[S_k^-(t) - S_{k'}^-(t) \right], \tag{24}$$

where $S^2 = (S_l^x)^2 + (S_l^y)^2$

Then the approximate system of equations defining the dispersion of the model spin system will be as follows:

$$[\{2S\sum_k [J - \hbar\omega]S_k^- - 2S\sum_{k'} JS_{k'}^- = 0. \tag{25}$$

This conceptual analysis allows us to imagine the complexity of the theoretical description of spin transport in the considered FM-CNT-FM systems, and the amount of computational work.

5.3 SPINTRONIC DEVICES DESCRIPTIONS

The first two-terminal CNT spin valve device was presented in [86], where spin-dependent transport was demonstrated for MWCNTs with 9% magnetoresistance ratio (MR) at 4.2 K.

The device shown in Figure 17 is the so-called CPP (current perpendicular to the plane) geometry. The resistance of such geometry is very low and difficult to detect. For practical applications, structures with the CIP (current in the plane) are used because they have higher resistance and thus higher difference with the magnetic field [78].

The GMR can be understood through the Mott's two-current model [83]. According to that model, the electrical conductivity of metal can be described by two more or less independent channels, one for majority spins and the other for minority spins. Scattering processes that conserve spin states are much more probable than the processes that flip spins. Another view proposed by Mott is that the scattering probability of spin up and spin down is quite different, independent on the nature of the scattering process [84]. This is shown schematically in Figure 17.

The difference in the resistance of ferromagnets can be explained by the exchange split band structure. The scattering of the electrons depends on where the electron band crosses the Fermi [78].

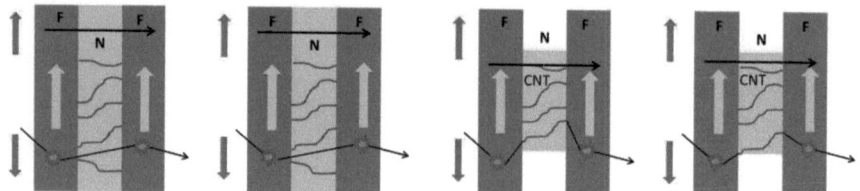

Figure 17 Giant magnetoresistance (GMR) device, a thin normal metal spacer (N) separates two ferromagnets (F), the current flows perpendicular to the plane of the sample, N space can be filled by introduced (or grown) CNT, e.g. – metal-like.

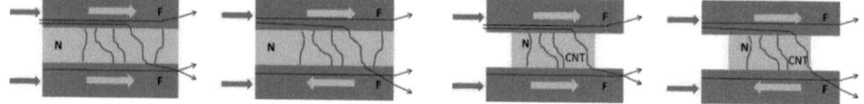

Figure 18 Current in Plane (CIP) spin-valve and equivalent resistor mode, separate channels are for minority and majority spins, the electrons scatter from one F layer to the other on the way through the sandwich.

As mentioned above, commercial spin-valves have normally the (CIP) (current in plane) geometry. This gives higher resistance and thus higher resistance difference between parallel and antiparallel spin orientations. Having the current in plane of the sandwich will qualitatively give the same effect as in CPP devices. This situation is shown schematically in Figure 18. Electrons with minority and majority spins are treated separately. When the electrons flow through the sandwich they will scatter back and forth from the upper F layer to the lower one.

We consider a possibility for the creation of magnetic memory devices on the magneto-resistance phenomenon (GMR or TMR) consisting of sandwiches of two ferromagnetic metals separated by a thin spacer layer of **normal metal** (see, e.g., Figure 18) **or semiconductor.** Spin-valves are of great industrial importance and are used as magnetic field sensors. The resistance of the device is dependent on the relative magnetization orientation of the ferromagnets. The TMR signal operates in the same way as the GMR, where R_P and R_A are the resistances of the device for parallel and antiparallel orientations, respectively, of the ferromagnets magnetization (Figures 17,18) [89]. Our idea is to reach the same effect by introducing metal or semiconductor-like CNTs into the N space after magnetically controlled CVD CNTs growth [89].

6 Conclusions

Electromagnetic properties of Pt- and Fe-CNT interconnects are considered from the point of view of mechanical stability and electrical efficiency. CNT-Fe interconnects are stronger mechanically. However, CNT-Pt interconnects having smaller resistances are more suitable electrically and more effective for various electronic nanodevices.

The use of Fe-Pt nanoclusters during the CVD magnetically stimulated growth allows controlling with some limitations in the diameter and chirality of growing CNTs. The diameter of nanoclusters defines the diameter of CNTs, while the angles of chiralities are correlated with the orientation of the external permanent magnetic field.

Fe-Pt nanoclusters and CNTs composition is a prototype of magnetic nanomemory devices. The magnetic efficiency depends on the stoichiometry coefficient 'x' and the ordering of Fe-Pt atomic structure.

Chemically ordered Fe-Pt nanoparticles, where we meet the sequence of Pt and Fe layers, allow us to provide all possible predicted magnetic properties. In case of substitutional disorder, this advantage disappears and we should talk about the percolation phenomenon of Fe_xPt_{1-x} ferromagnetism via 'x' alteration.

The CVD process of CNTs growth with the presence of Fe-Pt nanoparticles in the conditions of a strong magnetic field is a more orderly process.

Statistical dispersion of CNTs output according to the diameter is determined by a discrete number of effective bonds (especially in the case of small diameters of growing CNTs).

The dependence on the number of the effective bonds formed on the nanoparticle perimeter and the chirality angle is the essential tool for the morphology control of the future CNT.

The balance factor of the thermal energy and magnetic energy in deposited nanoparticles of carbon atoms plays the decisive role in evaluating the dispersion of chirality angles in CNTs.

The morphology of the forthcoming growing CNT stimulated by the magnetic field is set when forming the perimeter of the CNT base and the character of the magnetic orientation in effective bonds.

We can consider a CNT forest magnetically stimulated by the CVD process as a fragment of the magnetic nanomemory. CNT chirality in this case defines the efficiency of magnetic cells access (rates of exchange).

Magnetoresistance phenomenon (GMR and TMR) for nanomemory devices based on CNTs of various morphologies (i.e. various chiralities, diameters) including metal- and semiconductor-like ones, can be potentially viewed as an alternative for electromagnetic nanosensing and magnetic nanomemory.

Acknowledgments

This study was supported by Grant EU FP7 CACOMEL project FP7-247007, Call ID 'FP7-PEOPLE-2009-IRSES', 2010-2014 *Nanocarbon based components and materials for high frequency electronics and* Grant 0168/GF4 from the Ministry of Education and Science of the Republic of Kazakhstan. Evaluating the contribution of the research team members we should point out that Victor Gopeyenko and Nataly Burlutskaya made a large amount of calculations, Yuri Zhukovskii, Federico Micciulla and Aldo Capobianchi made the critical analysis of CVD details of CNTs growth, Yuri Shunin, Tamara Lobanova-Shunina, and Stefano Bellucci made the review of theoretical approaches and wrote most of the manuscript. We also thank Prof. E A Kotomin and Prof. S A Maksimenko for stimulating discussions on the topic.

References

[1] *Nanodevices and Nanomaterials for Ecological Security* 2012 Eds Shunin Yu and Kiv A Series: NATO Science for Peace Series B - Physics and Biophysics Hiedelberg: Springer Verlag 363

[2] Shunin Yu N, Zhukovskii Yu F, Burlutskaya N Yu, Gopejenko V I, Bellucci S 2012. *Nanodevices and Nanomaterials for Ecological Security* Series: NATO Science for Peace Series B - Physics and Biophysics Eds Yu Shunin and A Kiv Heidelberg: Springer Verlag, 237-62

[3] Shunin Yu N, Zhukovskii Yu F, Gopejenko V I, Burlutskaya N, Lobanova-Shunina T and Bellucci S 2012 *Journal of Nanophotonics* 6(1) 061706-1-16

[4] Shunin Yu N, and Schwartz K K 1997 *Computer Modelling of Electronic and Atomic Processes in Solids* Eds R C Tennyson and A Kiv Dodrecht/Boston/London: Kluwer Acad. Publisher 241-57

[5] Shunin Yu N, Zhukovskii Yu F, Gopejenko V I, Burlutskaya N and Bellucci S 2011 *Nanoscience and Nanotechnology Letters* 3(6) 816-825

[6] Economou, E L 2006 *Green's Functions in Quantum Physics Solid State* Ser. 7 Berlin/Heidelberg: Springer Verlag

[7] Ziman J M 1979 *Models of Disorder* New York-London: Cambridge Univ. Press Chapter 10

[8] Shunin, Yu N, Zhukovskii Yu F, Burlutskaya N, Bellucci S. 2012 *Journal of Nanoelectronics and Optoelectronics* 7(1) 3-11

[9] Capobianchi A, Colapietro M, Fiorani D, Foglia S, Imperatori P, Laureti S, and et al 2009. *Chem. Mater.* 21(10) 2007-9

[10] Capobianchi A, Campi, G, Camalli M, and Veroli, C Z 2009 *Kristallogr* 224(8) 384-8

[11] Faustini M, Capobianchi A, Varvaro G, and Grosso D 2012 *Chem. Mater.* 24 1072-9

[12] Bellucci S 2005 *physica status solidi* (c) 2(1) 34-47

[13] Moisala A, Nasibulin A G, and Kauppinen E I J 2003 *Phys. Condens. Mater.* 15 S3011-35

[14] Baker R T K, Hams P S 1978 *Chemistry and Physics of Carbon* 14 Eds P L Walker Jr and P A Thrower New York: Marcel Dekker 83

[15] Muller T E, Reid D G, Hsu W K, Hare J P, Kroto H W, Walton D R M 1997 *Carbon* 35(7) 951-966

[16] *The Science and Technology of Carbon Nanotubes* 1999 Eds K Tanaka, T Yamabe, K Fukui Amsterdam: Elsevier Science Ltd

[17] Sen R, Govindaraj A, Rao C N R 1997 *Chem. Phys. Lett.* 267(3) 276-80

[18] Satishkumar B C, Govindraj A, Sen R, and Rao C N R 1998 *Chem. Phys. Lett.* 293(1) 47-52

[19] *Carbon Nanotubes Science and Applications* 2005 Ed Meyyappan Florida: CRC Press LLC Google e-Book

[20] Xie S., Chang B, Li W, Pan Z, Sun L, Mao J, and et al. 1999 *Adv. Mater.* 11 1135-9

[21] Cassell A M, Raymakers J A, Kong J, and Dai H J 1999 *Phys. Chem.* B **103** 6484-9

[22] Su M, Zheng B, Liu J 2000 *Chem. Phys. Lett.* **322** 32-6

[23] Su M, Li Y, Maynor B, Buldam A, Lu J P, Liu J J 2000 *Phys. Chem.* B **104** 6505-12.

[24] Fan S S, Chapline M G, Franklin N R, Tombler T W, Cassell A M, Dai H 1999 *Science* **283** 512-4

[25] Cheng H M, Li F, Su G, Pan H Y, He L L, Sun X et al. 1998 *Appl. Phys. Lett.* **72** 3282-4

[26] Li Y, Liu J 2001 *Chem. Mater.* **13** 1008-14

[27] Lee D C, Mikulec F V, Korgel B A 2004 *Amer. Chem. Soc.* **126**(15) 4951-7

[28] Chen G Y, Jensen B, Stolojan V, Silva S R P 2011 *Carbon* **49**(1) 280-5

[29] Shang N G, Tan Y Y, Stolojan V, Papakonstantinou P, Silva S R P 2010 *Nanotechnology* **21**(50) 505604

[30] Kumar M, Ando Yo 2010 *Journal of Nanoscience and Nanotechnology* **10**(6) 3739-58

[31] Takagi D, Hibino H, Suzuki S, Kobayashi Y, Homma Y 2007 *Nano Lett.* **7** 2272-5

[32] Takagi D, Homma Y, Hibino H, Suzuki S, Kobayashi Y 2006 *Nano Lett.* **6** 2642-5

[33] Liu H, Takagi D, Ohno H, Chiashi Sh, Chokan T, Homma Yo 2008 *Appl. Phys. Express* **1** 014001-3

[34] Hernadi K, Fonseca A, Nagy J B, Bernaerts D, Lucas A A, 1996 *Carbon* **34**(10) 1249-57

[35] Kong J, Cassell A M, Dai H 1998 *Chem. Phys. Lett.* **292**(4) 567-74

[36] Li W Z, Xie S, Qian L X, Chang B H, Zou B S, Zhou W Y, Zhao R A, Wang G 1996 *Science* **274** 1701-3

[37] Wei B Q, Vajtai R , Jung Y , Ward J , Zhang R , Ramanath G , Ajayan P M 2002 *Nature* **416** 495-6

[38] Nikolaev P, Bronikowski M J , Bradley R K , Rohmund F , Colbert D T, Smith K A, Smalley R E 1999 *Chem. Phys. Lett.* **313** 91-7

[39] Dai H, Rinzler A G, Nikolaev P, Thess A, Colbert D. T. and Smalley R E 1996 *Chem. Phys. Lett.* **260** 471-5

[40] Cheng H M, Li F, Sun X., Brown S D M., Pimenta M A., Marucci A, Dresselhaus G, Dresselhaus M S 1998 *Chem. Phys. Lett.* **289** 602-10

[41] Hafner J H, Bronikowski M J , Azamian B R, Nikolaev P, Rinzler A G , Colbert D T , Smith K A, Smalley R E 1998 *Chem. Phys. Lett.* **296** 195-202

[42] Flahaut E, Govindaraj A, Peigney A, Laurent C, Rao C N R 1999 *Chem. Phys. Lett.* **300**(1-2) 236-42

[43] Gruneis A, Rummeli M H, Kramberger C, Grimm D, Gemming T, Barreiro A, Ayala P, Pichler T, Kuzmany H, Schamann C, Pfeiffer R, Schumann J, Buchner B 2006 *physica status solidi* (b) **243** 3054-7

[44] Maruyama S, Miyauchi Y, Edamura T, Igarashi Y, Chiashi S and Murakami Y 2003 *Chem. Phys. Lett.* **375** 553-9

[45] Maruyama S, Kojima R, Miyauchi Y, Chiashi S, Kohno M, 2002 *Chem.Phys.Lett.* **360**(3) 229-34

[46] Okubo S, Sekine T, Suzuki S, Achiba Y, Tsukagoshi K, Aoyagi Y and Kataura H, 2004 *Jpn. J. Appl. Phys.* **43** L396-8

[47] Gruneis A, Rummeli M H, Kramberger C, Barreiro A, Pichler T, Pfeiffer R, Kuzmany H, Gemming T, Buchner B 2006 *Carbon* **44** 3177-82

[48] Nasibulin A G, Moisala A, Jiang H, Kauppinen E I 2006 *J. Nanopart. Res.* **8** 465-75

[49] Murakami Y, Miyauchi Y, Chiashi S, Maruyama S 2003 *Chem. Phys. Lett.* **377**(1) 49-54

[50] Murakami Y, Chiashi S, Miyauchi Y, Hu M, Ogura M, Okubo T, Maruyama S 2004 *Chem. Phys. Lett.* **385**(1) 298-303

[51] Maruyama S, Einarsson E, Murakami Y, Edamura T 2005 *Chem. Phys. Lett.* **403**(4-6) 320-3

[52] Xiang R, Einarsson E, Okawa J, Miyauchi Y and Maruyama S 2009 *J.Phys.Chem.* C **113** 7511-5

[53] Yuan D, Ding L, Chu H, Feng Y, McNicholas T P, Liu J 2008 *Nano Lett.* **8**(8) 2576-9

[54] Ward J, Wei B Q, Ajayan P.M. 2003 *Chem.Phys.Lett.* **376**(5-6) 717-25

[55] Morjan R E, Nerushev O A, Sveningsson M, Rohmund F, Falk L K L, Campbell E E B 2004 *Appl. Phys.* A **78**(3) 253-61

[56] Ago H, Komatsu T, Ohshima S, Kuriki Y, Yumura M 2000 *Appl.Phys.Lett.* **77**(1) 79-81

[57] Kumar M, Ando Y 2005 *Carbon* **43**(3) 533-40

[58] Li W Z, Wen J G, Ren Z F 2002 *Appl. Phys.* A **74**(3) 397-402

[59] Li W Z, Wen J G,, Tu Y, Ren Z F 2001 *Appl. Phys.* A **73** 259-254

[60] Maruyama S, Murakami Y, Shibuta Y, Miyauchi Y, Chiashi S 2004 *J. Nanosci. Nanotechnol.* **4** 360-7

[61] Kumar M and Ando Y 2008 *Defence Sci. Journal* **58**(4) 496-503

[62] Raty J-Y, Gygi F, Galli G 2005 *Phys.Rev.Lett.* **95** 096103-7

[63] Shunin Yu N, Zhukovskii Yu F, Bellucci S 2008 *Comput. Model. & New Technol.* **12**(2) 66-77

[64] Shunin Yu N 1991 *Simulation of atomic and electronic structures of disordered semiconductors* Dr.Sc.Habil. Thesis (Phys.&Math.) Riga-Salaspils

[65] Shunin Yu N, Zhukovskii Yu F, Burlutskaya N Yu, Bellucci S 2011 *Central European Journal of Physics* **9**(2) 519-29

[66] Gaspar R 1952 *Acta Phys. Acad. Sci. Hung.* **2** 151-78 ;1954 *Acta Phys. Acad. Sci. Hung.***3** 263-86

[67] Slater J C 1974 *The Self-Consistent Field for Molecules and Solids* Vol.**4** McGraw-Hill Book Company, N.-Y.

[68] Ehrenreich H, Schwartz L 1976 *The Electronic Structure of Alloys* Sol .St. Phys. Vol. **31**, Academic Press, N.-Y.-San Francisco-London

[69] Bellucci S., Zhukovskii Yu F, Gopejenko V I, Burlutskaya N, Shunin Yu N 2012 *CIMTEC2012 Theses* June 10-14 Montecatini Terme Tuscany Italy 78

[70] Sun A C, Kuo P C, Chen S C, Chou C Y, Huang H L, Hsu J H 2004 *J. Appl. Phys.* **95**(11) 7264-6

[71] Elkins K, Li D, Poudyal N, Nandwana V, Jin Zh, Chen K, Liu J P 2005 *J. Phys.* D **38** 2306-9

[72] Yan M L, Sabirianov R F , Xu Y F , Li X Z , Sellmyer D J 2004 *IEEE Trans. on Magnetics* **40**(4) 2470-2

[73] Kim Y-H , Choi J, Chang K J 2003 *Phys. Rev.* B **68** 125420

[74] Medwal R., Sehdev N, Annapoorni S J 2012 Phys. D *Appl. Phys.* **45** 055001-7

[75] Arabshahi H, Hematabadi A, Bakhshayeshi A, Ghazi M 2012 *World Applied Programming* **2**(8) 415-20

[76] Baibich M N, Broto J M, Fert A, van Dau F N, Petroff, F, Eitenne P, and et al 1988 *Phys. Rev. Lett.* **61** 2472-5

[77] Binasch G, Grünberg P, Saurenbach F, Zinn W 1989 *Phys. Rev.* B **39** 4828-30

[78] Tsymbal E Y, Pettifor D G 2001 *Solid State Physics* **56** 113-237

[79] Julliére M 1975 *Phys. Lett.* A **54** 225-6

[80] Moodera J S, Kinder L R, Wong T M, and Meservey R 1995 *Phys. Rev. Lett.* **74** 3273-6

[81] Parkin S S P, More N, and Roche K P 1990 *Phys. Rev. Lett.* **64** 2304-7

[82] Parkin S S P, Bhadra R, Roche K P 1991 *Phys. Rev. Lett.* **66** 2152-5

[83] Mott N F 1936 *Proc. Royal. Soc.* **153** 699-726

[84] Mott N F 1936 *Proc. Royal. Soc.* **156** 368-82

[85] A. Barth´el´emy, A. Fert, J.-P. Contour, M. Bowen, V. Cros, J. M. De Teresa, A. Hamzic, J. C. Faini, J. M. George, J. Grollier, F. Montaigne, F. Pailloux, F. Petroff, and C. Vouille 2002 *Journal of Magnetism and Magnetic Materials* **242–245**(1) 68–76

[86] Tsukagoshi K, Alphenaar B W, Ago H. 1999 *Nature* **401**(6753)572–574

[87] Hewson A C 1993 *The Kondo problem to heavy fermions* Cambridge: Cambridge University Press

[88] Céspedes O, Ferreira M S, Sanvito S, Kociak M, Coey J M D 2004 *Journal of Physics: Condensed Matter* **16**(10) L155-61

[89] Shunin Yu N, Zhukovskii Yu F, Gopeyenko V I, Burlutskaya N Yu, Lobanova-Shunina T D, Bellucci S 2015 *Proc Int Conf NANOMEETING-2015 PHYSICS, CHEMISTRY AND APPLICATION OF NANOSTRUCTURES, 2015* Eds V E Borisenko, S V Gaponenko, V S Gurin, C H Kam New-Jersey-London-Singapore-Beijing-Hong Kong-Taipei, Chennai: World Scientific 2015 207-10

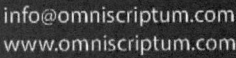